YOUR KNOWLEDGE HAS VALUE

- We will publish your bachelor's and
  master's thesis, essays and papers

- Your own eBook and book -
  sold worldwide in all relevant shops

- Earn money with each sale

Upload your text at www.GRIN.com
and publish for free

Neha Khandelwal

Protease Inhibitors: Potential and Constraints

Biopesticides

GRIN Verlag

Bibliografische Information der Deutschen Nationalbibliothek:

Die Deutsche Bibliothek verzeichnet diese Publikation in der Deutschen National-
bibliografie; detaillierte bibliografische Daten sind im Internet über http://dnb.d-
nb.de/ abrufbar.

Dieses Werk sowie alle darin enthaltenen einzelnen Beiträge und Abbildungen
sind urheberrechtlich geschützt. Jede Verwertung, die nicht ausdrücklich vom
Urheberrechtsschutz zugelassen ist, bedarf der vorherigen Zustimmung des Verla-
ges. Das gilt insbesondere für Vervielfältigungen, Bearbeitungen, Übersetzungen,
Mikroverfilmungen, Auswertungen durch Datenbanken und für die Einspeicherung
und Verarbeitung in elektronische Systeme. Alle Rechte, auch die des auszugsweisen
Nachdrucks, der fotomechanischen Wiedergabe (einschließlich Mikrokopie) sowie
der Auswertung durch Datenbanken oder ähnliche Einrichtungen, vorbehalten.

Imprint:

Copyright © 2011 GRIN Verlag GmbH
Druck und Bindung: Books on Demand GmbH, Norderstedt Germany
ISBN: 978-3-656-37898-3

This book at GRIN:

http://www.grin.com/en/e-book/209929/protease-inhibitors-potential-and-constraints

GRIN - Your knowledge has value

Der GRIN Verlag publiziert seit 1998 wissenschaftliche Arbeiten von Studenten, Hochschullehrern und anderen Akademikern als eBook und gedrucktes Buch. Die Verlagswebsite www.grin.com ist die ideale Plattform zur Veröffentlichung von Hausarbeiten, Abschlussarbeiten, wissenschaftlichen Aufsätzen, Dissertationen und Fachbüchern.

Visit us on the internet:

http://www.grin.com/

http://www.facebook.com/grincom

http://www.twitter.com/grin_com

Protease inhibitors as biopesticides: Potential and constraints

Neha Khandelwal, Rakesh S. Joshi, Vidya S. Gupta and Ashok P. Giri*

Plant Molecular Biology Unit, Division of Biochemical Sciences,
National Chemical Laboratory (CSIR)

Corresponding Author: Ashok P. Giri

Abstract: Plant protease inhibitors are the well studied class of plant defensive proteins. High level of up-regulation upon insect damage, significantly elevated levels in reproductive and storage organs as well as specificity to pest proteases ascertain their defensive function. In the present chapter, we introduced molecular basis of plant-insect interactions and focused on different classes of protease inhibitors based on their target protease(s). In particular serine protease inhibitors are well-studied in plants and the effort to deploy them for plant protection methods has been discussed. Wound-induced serine protease inhibitors Pin-I and Pin-II type are in major focus due their unique features such as high binding specificity to target proteases, multi-domain structure, stability and processing in digestive tract of insects. Numerous efforts by different research groups all over the world for their application for insect control have been discussed emphasizing the merit and constraints. It is known that insect respond to protease inhibitors by expressing variants of insensitive proteases which nullify the effect of protease inhibitors. However, there are several unanswered questions that needs to be addressed for the successful use of proteinase inhibitors (i) What are the protein engineering and molecular biology approaches for their competent use against insect pests? (ii) How different strategies can be applied for the successful implementation of protease inhibitors commercially? (iii) How to explore selective use of broad spectrum or specific protease inhibitors? etc. These issues are discussed throughout the review by emphasizing more on limitations of protease inhibitors as bio-pesticide agents to date.

Keywords: Biopesticides, Lepidopteran insects, Protease inhibitors and Transgenic plants

Insect pests: A major challenge in sustainable agriculture

Agriculture production in India significantly plays major role in country's economic development by accounting over 28% of largest global pulse production. India is the second largest producer of vegetables, cotton, wheat and rice. Agriculture in India has gone through immense transitions ranges from low productivity due to pest attack and high productivity by the use of chemical pesticides. About 3% of the total pesticides used in the world are utilized in India. The used pattern of pesticides reflects that cotton crop alone consumes 44.5% pesticides estimating USD 250 million for the control of bollworms which is followed by rice, accounting 22.8% of pesticide consumption for the control of yellow stem borer. These two crops consume more than two thirds of the total quantity of the pesticides used in the country (Report by Dileep K. Singh). Other key pests of similar importance are stem borers of sorghum and maize, fruit and shoot borer of brinjal, fruit and pod borer of tomato and chickpea and diamond back moth of cruciferous crops, cabbage and cauliflower. These pests are perennial and persistently causing losses to these economically important crops (Reddy and Zehr, 2004) An average estimation of the damage caused by insects throughout the year has been shown in terms of percentages **(figure 1)**. Past fifty years the usage of chemical insecticides and pesticides is at their peak for pest control in various crop species. Around 50 per cent of the total insecticide consumption is that of organophosphates, followed by the synthetic pyrethroids (19%), organochlorines (18%), carbamates (4%) and biopesticides (1%). Most of the chemical pesticides become detrimental to human health as well as raises environmental concern when used above threshold limit and results in eradication of beneficial insects. In addition, most of the agronomically important pests have developed resistance towards it which causes the use of such insecticides inefficient at moderate level and demands for their increased use on agricultural crops. Sustainable agricultural practices are thus of global concern today which takes maximum utilization of environment without causing any harm to it.

Despite the use of various protective measures, pests have remained as a potential threat to agricultural crops. Insects that cause serious economic loss belong to lepidopteran, coleopteran, hemipteran, orthopteran and dipteran families. As pollinators of many plants, lepidopteran adult moths and butterflies are usually beneficial insects that feed on nectar. Caterpillar however consists of chewing mouth parts which allow them to feed on plant parts causing extensive damage. Unlike lepidopteran, coleopteran beetles have been known to cause damage usually in their adult stage, for instance *Caryedon serratus* (Olivier) and *Phyllotreta cruciferae* infects groundnut and seedling phase of cabbage (Ranga Rao *et al.*, 2010; K Mayuri and G. Mikunthan, 2009). Much of the emphasis has been given on lepidopterans which represent the most damaging class of insects that feeds on cereals, pulses, vegetables, and oil seeds. Research in last few decades are much focused in studying about the interaction between plants and insects to develop plant protection strategy against them. Crop biotechnology, which broadly includes areas of development of transgenic crops, structural and functional genomics and marker-assisted breeding, could provide us with the vital breakthroughs to achieve improvements in both quality and quantity in a sustainable manner (Manju Sharma *et al.*, 2003).

Plant-insect interaction

In natural ecosystem, plants and insects are in constant interaction with each other. This complex and dynamic interaction between plants and insects are being studied from molecular, ecological and evolutionary perspective. Insects, as pollinators and predators of undesirable pests provide beneficial role, though many insects share same life cycle (**figure 2**), but many have emerged as harmful due to their feeding habit on enormous number of plant species. The number of plant attackers is seemed to be almost limitless whose knowledge is essential in respect to develop effective strategy for the protection of crop plants. Insects that feed on plants are categorized into monophagous, oligophagous and polyphagous depending upon the number of species or families they feed upon.

A. **Monophagous or specialized pests**- Insects which are known to feed on single species of plant and during the absence of their main host plants they shift to some other related host species temporarily. Examples includes Pink bollworm (*Pectinophora gossypiella*), rice stem borer (*Scirpophaga incertulas*), brinjal shoot and fruit borer (*Leucinodes orbonalis*) (Hanur, 2008)

B. **Oligophagous pests**- Insects that confine their feeding activity belonging to one family, a well known example is cabbage diamondback moth (*Plutella xylostella* L.; Plutellidae, Lepidoptera)

C. **Polyphagous or generalized pests**- feeds on various species of plants belonging to more than one taxonomic group or family e.g. cotton bollworm or pod borer (*Helicoverpa armigera*), Tobacco cut worm (*Spodoptera litura*)

Polyphagous insects are considered as the most threatening among above categories due to their adaptation to various host plants posing problem in controlling them by employing same strategy with different crops. *Spodoptera exigua* for example feeds on approximately 50 plant species including pigweed, *Amaranthus hybridus*, and cotton, *Gossypium hirsutum* (Showler *et al.,* 2001). Similarly *Heliothis virescens* mostly feed on cotton, soybean, flax, alfalfa but found at high density on tobacco being it the most preferable crop (Tilmann, 2006). Polyphagous pests possess the ability to adapt their digestive system in response to diverse type of host plant they feed upon, by switching on/ off genes involved in the metabolism of the host plant components. The well known example is the *Helicoverpa armigera* (Lepidoptera: Noctuidae) which has been widely studied for its adaptability on various host plants. Voracious caterpillars of *H. armigera* feeds on approximately 200 plant species including number of forest and fruit trees along with agronomically important crops such as cotton, chickpea, pigeon pea, tomato, maize, okra etc. by metabolizing nutrients for their growth and development. In order to grow on varied type of host plants these insects bring about molecular changes in their gut complement (Patankar et al, 2001). These molecular flexibilities enable insect's digestive system to react efficiently upon exposure to different crops by synthesizing more isoforms of proteases, if feeding on protein rich diet and amylases if feeding on starch rich diet. Thus insects are capable of surviving on enormous number of plants due to which it has attained the status of notorious pest. Several studies have been attempted in order to control *H. armigera,* which alone is responsible for causing heavy losses of economically important crops ranging from minimum 14% to maximum loss of 100%. Currently, control of insect pests is mainly dependent on the use of chemical insecticides which are commercially available in large numbers. Insect pest management by chemicals obviously has brought about considerable protection to crop yields over the past five decades. Pesticides such as lambda cyhalothrin, cypermethrin, and quinalphos are amongst few chemicals which have been formulated and utilized in pest management. Though they have been found to be efficient in reducing insect proliferation, their detrimental effect on ecosystem has to be considered. In search of environmentally benign approach transgenic plants have been emerged as an efficient strategy for the pest control management. Bt transgenics were successfully implemented for the control of major field pest, its implications and constrains are discussed partly in the following sections.

Plant defense system against insects

Since years of interaction plants and insects, both have coevolved in order to survive in their changing niches. Insects gradually adapt to take maximum benefit from the host while plants on the other hand have evolved to protect themselves. Being sedentary in nature plants have developed the ability to react spontaneously against insect attack by synthesizing number of defensive molecules. These responses of plants have been extensively studied to exploit them for crop protective measures. Plants show different types of defence responses against pathogens and pests which are categorized into direct and indirect responses (**figure 3**). Direct mode includes production of toxic, repellent or anti digestive molecules like cyanogenic glucosides, glucosinolates, alkaloids, phenolics, proteinase inhibitors etc, along with this they also pose some physical barriers (e.g. Cuticle, Trichomes, Thorns etc.) which play a role in primary protection against infestation (Bennett *et al.*, 1994). In indirect mode, plant emits some volatiles which produce the extra floral nectar that attracts predators of insect herbivorous to gain indirect means of protection (Kessler and Baldwin, 2001; 2002).

Wounding by insect herbivory switches on different signalling reflexes in plants, leads to induction of different defence mechanisms (**figure 4**). When an insect feeds on plant, some elicitor molecules are transferred through the saliva of insect. These elicitor molecules cause local secretion of some peptide hormones like prosystemin, threonin deaminase etc. (Ferry *et al.*, 2004; Kang *et al.*, 2006). Prosystemin undergoes proteolytic cleavage to produce active systemin. Production of systemin switches the Octadeconide (OD) pathway, which catalyzes the breakdown of linolenic acid and the formation of jasmonic acid (JA) (Ferry *et al.*, 2004). Jasmonic acid reacts with free isoleucine, which is produce active signalling molecule. On cleavage threonine gives isoleucine (react with jasmonic acid to gives JA-Ile) and α keto butyrate (Toxic to insect) this reaction is catalyzed by threonin deaminase (Kang *et al.*, 2006). JA-Ile switches on early genes in vascular bundle which cause activation of NADPH oxidase. NADPH oxidase catalyzes the production of hydrogen peroxide (H_2O_2) (Orozco-Cardenas M and Ryan, 1999). H_2O_2 is transfer to mesophyll cells, where they activate late gene responsible for production of defence molecule like proteinase inhibitor (PIs). In other way of defence Jasmonic acid induces the release of some volatiles that attract natural predators of the insect pest.

Jasmonic acid on methylation produces methyl jasmonate. Methyl Jasmonate acts as a mediator molecule to give plant-plant interaction (Orozco-Cardenas M *et al.*, 2001; Rojo *et al.*, 2003).

Role of protease inhibitors in plants

PIs are of common occurrence in the plant kingdom and initially were found to be abundant in storage tissues such as tubers and seeds. They protect storage organs from predators by inhibiting hydrolytic activity of target proteases and play developmental role by the regulation of protein turnover. Later on their presence has also been detected in the aerial parts of plants where their expression is induced in response to pest attack (**figure 5**). Protease inhibitors mostly have been characterized from Fabaceae (legumes), Poaceae (cereals) and Solanaceae families. Since last decade insects digestive proteolytic enzymes have come under intensive investigation thus major attention has been given in exploitation of defensive role of protease inhibitors which induced upon insect attack for the purpose of crop protection. Thorough understanding of protease inhibitors from its induction to its inhibitory effect on insects has been determined in various studies to characterize their potential role in controlling pest population. PIs are responsible in causing reduction in the availability of amino acids by blocking the hydrolytic activity of proteases and were first shown as plant defensive proteins in 1972 when their induction in potato and tomato was observed due to wounding and insect herbivory. Subsequently Gatehouse et al and co-workers demonstrated the resistance of a cowpea variety to a bruchid beetle (Coleoptera) due to the elevated trypsin inhibitor (Donald Boutler *et al.*, 1990). PIs act on insects by interfering with the metabolic breakdown of the protein in the insect midgut thereby arresting their catabolism that leads to the accumulation of the protein causing physiological burden on insects. Most of the lepidopteran insects prefers protein rich diet hence employment of protease inhibitor have been proven to be the most promising approach for the development of effective strategy against insect pest. Natural mutants of trypsin inhibitors in wild tobacco plants recently provided the first true evidence that the inhibitors exert a certain level of control even against their natural 'adapted' pests (Jongsma and Beekwilder, 2008). In addition to resistance from insects, PIs are also known to confer resistance from abiotic stresses such as salt stress, change in pH and osmolarity (Tamura *et al.*, 2003). Strategically use of protease inhibitors in integrated crop management requires the knowledge of the synthesis and role of PIs in plant system in response to insect attack.

Mode of action of protease inhibitor against insect proteases

The different classes of inhibitors are distinguished by the structure of their polypeptide backbone. However the mechanism of binding of the plant protease inhibitors to the insect proteases appears to be similar with all the four classes of inhibitors. Interaction between enzyme and inhibitor occurs in canonical (substrate like) manner. The inhibitor binds to the active site on the enzyme to form a complex with a very low dissociation constant (10^7 to 10^{14} M at neutral pH values), thus effectively blocking the active site. For serine proteinase inhibitors, most residues interacting with the proteases are located on a single loop, to which the P_1 residue is central. The P_1 residue of protease substrates is the one on the amino-side of the hydrolyzed bond, and is often crucial for substrate recognition. The surrounding residues (P_2, P_3 etc. at its N-terminus, and P_1', P_2' etc. at its C-terminus) play secondary roles in the interaction of proteases and their substrates. Both in inhibitors and in substrates, the P_1 residue fits into the S_1 substrate binding site of the proteinase. Unlike normal peptide substrates, the inhibitor residues around P_1 interact with the enzyme through complementary polar and hydrophobic interactions, and are held in position by strong bonds with the inhibitor scaffold. This prevents immediate dissociation of the complex, thus keeping the enzyme inactive. P_1 residue in reactive loop of inhibitor is primary determinant of inhibitor specificity. If P_1 residue is positively charged like Arg and Lys they are act as trypsin inhibitor. In case of chymotrypsin inhibitor P_1 is hydrophobic in nature (Leu, Ile, Trp or Phe). The inhibitor residues around P_1 residue are from gene family which more hypervariable in nature relative to most of protein.

Distinct classes of protease inhibitors

Protease inhibitors are categorized into four classes on the basis of the mechanistic classes of the proteases they inhibit: serine PIs, cysteine PIs, asparatate PIs and metallo proteases PIs (**figure 5**). Serine PIs have a widespread distribution in plant kingdom and are extensively studied as most of the lepidopteran insects utilize active serine proteases for their nutrient requirement by metabolic breakdown of the protein.

Cysteine PIs represent the second most well studied class of protease inhibitors whereas little is known about asparatate and metallo protease inhibitors.

Serine protease inhibitors

Plant serine PIs acts against proteases like trypsin, chymotrypsin, elastases and subtilisin (bacterial enzyme). Trypsin- and chymotrypsin-like enzymes are predominantly present in the midgut of lepidopteran insects. These enzymes are active in the alkaline pH range from 9 to 11 of gut proteolytic environment. Protease function of cleaving proteins at specific sites, enable insects to utilize them for their nutrient enrichment. Trypsin cleaves at the C-terminal residue of lysine or arginine containing basic side chains whereas chymotrypsin cuts at the C-terminal residue of phenylalanine, trypsin and leucine, all containing hydrophobic side chain. Various researchers have identified the presence of serine protease inhibitors in plants for their defensive role and divided into various families of which four major types are: Kunitz, Bowman birk, squash and wound inducible type. Latter is further categorized in to two families namely Potato inhibitor type 1 (Pin I) and potato inhibitor type 2 (Pin II). Different families of serine PIs were recognized based on their distribution in plant kingdom, sequence and functional divergence.

As wounding induces high accumulation of the PIs in plant parts, one might ask about their consequential effect on native proteinases of the plants. Major digestion of protein in plants occurs by the proteases which contain cysteine residues at their active site rather than serine residue thus the significant amount of serine PIs in plants does not affect plants metabolic function and cannot be used for regulating endogenous proteases. This clearly reflects the presence of different type of proteinase inhibitors for the regulation of protein turnover in plants. However few reports have emerged describing endogenous function of serine proteinase inhibitors in plants (Hartl *et al.,* 2010). Moreover, many plants express a number of different PIs and it was unknown if these proteins work synergistically as defenses or if they also have other functions. In case of serine PIs, gene- and tissue-specific expression patterns suggest multiple functions in generative tissues, including a possible involvement in development. Role of PIs have been characterized from various plants towards their defensive function against pests. From example, recently, protease inhibitors from *Capsicum annuum* plant were also accessed for their expression in different parts of plants. Single and double inhibitory domain proteins were found to be strongly expressed in stems, whereas multi IRD protein expression was restricted to leaves which clearly distinguish their role in defence from endogenous function (Tamhane et al, 2009). Role of plant serine PIs in defence has been well documented as they were shown to inhibit insect digestive enzymes leading to retardation in their growth and development. Broadway and Duffey in their studies have demonstrated the antinutritional effect of serine protease inhibitors which showed inhibition of *Heliothis Zea* and *Spodoptera exigua* after incorporated in artificial diet (Broadway and Duffer, 1986). In another similar study PIs of tobacco (*Nicotiana alata*) have shown to affect *H. punctigera* larval and Black field cricket (*Teleogryllus commodus)* growth (Heath *et al.*, 1997). Following this several serine protease inhibitors from non host plants were investigated for their potential in conferring resistance to plants and attempts were made to develop transgenic plants.

Cysteine Protease inhibitors

Cysteine protease inhibitors are also called plant cystatins or phytocystatins and are divided into two groups one group containing single inhibitory domain whereas second group includes PIs carrying multiple domains. Cysteine PIs have been characterized from several plants viz. cowpea, potato, cabbage, ragweed, carrot, papaya, apple fruit, avocado, chestnut and various crop plants like sunflower, rice, wheat, maize, soybean and sugarcane (Shu-Guo FAN and Guo-Jiang WU, 2005). Cysteine PIs acts as an antagonist for cysteine proteases which are active in the insect midgut of pH ranges from 5 to 7. Cysteine proteases are usually found in coleopteran order of insects such as Cowpea weevil (*C. maculates*), Bruchid *Zabrates subfaceatus*, Flour beetle (*Tribolium castaneum*), Mexican beetle (*Epilachna varivestis*) and the Bean weevil *Ascanthoscelides beetles*. Studies revealed that in few insects cysteine proteases also get inhibited by serine protease inhibitor in addition to cysteine PIs for eg. *C. maculatus* proteolytic enzymes get inhibited by Cowpea and soybean Trypsin inhibitor (serine PI) (Lawrence, 2002). It could be possible if they share high sequence homology between cysteine and serine protease inhibitors and thus cross interaction might occur. Oryzacystatin from rice have been well characterized among other cysteine PIs for its role in plant protection and attempts were also made to develop transgenic plants. Oryzacystatin contains highly conserved motifs glu-val-val-ala-gly (QVVAG) and pro-trp (PW) of which the former is the primary region which involve in interaction with the enzyme and PW motif acts like a cofactor (Arai *et al.*, 1991). Other than inhibitory effect on herbivory, cysteine proteases and their inhibitors are known to participate in program cell death according to the study conducted on soybean cells. Endogenous cysteine protease inhibitors may function in modifying PCD that is activated during oxidative stress and pathogen attack.

Asparatate and metallo protease inhibitors

Asparatyl and metallo protease inhibitors are the least studied class of protease inhibitors and thus have described here together. Asparatyl protease inhibitors are active against aspartate proteases which are usually found in hemipteran insects. These digestive enzymes display proteolytic activity at slightly acidic pH 3 to5. Asparatyl protease inhibitors have been found in sunflower, barley, cardoon flowers and also in potato tubers. An asparatyl PI, cathepsin D have considerable homology with Soyabean Trypsin inhibitor and thus inhibits trypsin and chymotrypsin along with asparatyl protease cathepsin D but does not inhibit other asparatyl proteases like rennin, cathepsin E and pepsin (Habib and Khalid, 2007). This suggests that specificity of protease inhibitors is sequence dependent. Small variation in sequence may result in changing the specificity and function of protease inhibitors. Such property can be exploited in order to design potent PI which will be efficient against not only one class of proteases but other classes as well. Insects are also known to modulate their digestive protease complement in response to asparatate inhibitors to compensate for the loss of protease activity (Graham *et al.*, 1997). Metalloprotease inhibitors are divided into two families, that inhibit metallo carboxypeptidases called metallocarboxypeptidase inhibitor which have been characterized from tomato and potato and other is cathepsin D inhibitor family from potatoes. Metallocarboxypeptidase inhibitors have been found to play developmental as well as protective role as their existence have been observed in potato tubers along with proteinase inhibitor 1 and 2. On the other hand they presence is also reported in potato leaf tissues as a response of wounding (Hollander-Czytko *et al.*, 1985; Liang and MacManus, 2002).

Families of serine protease inhibitors (SPIs)

Serine protease inhibitors are most well studied type of inhibitor. Such studies have provided a basic understanding of the mechanism of action that applies to most serine proteinase-inhibitor families. Depending upon amino acid composition, reactive loop confirmation and structure serine protease inhibitor are categorized into different families as follows.

1) Kunitz type protease inhibitor (KPI) family

Its occurrence is widespread in cereals, solanaceous species, and *Arabidopsis thaliana*, but they were first described in legumes (Liang and MacManus, 2002). They possess the ability to inhibit different proteases like trypsin, chymotrypsin, substilin, tissue plasminogen activator. Kunitz type protease inhibitors are also produced under stress condition, as has been found in potato tubers (*S. tuberosum*) and inhibits Cathespin D (Aspartic Proteinase) and papain (cysteine proteinase) (Ritonja *et al.*, 1990; Brzin *et al.*, 1988). Their molecular weight is around 18 to 22 KD and its structure consists of two disulfide bonds which reduces flexibility in reactive loop. The structure of Kunitz family inhibitor has been determined from several species. Soybean trypsin inhibitor is one of the well studied candidates; roughly it is spherical molecule with 12 antiparallel β strands. Reactive site is on protruding loop and the P1 carbonyl group remain in plane and forms a nominal trigonal geometry (Song and Suh, 1998). The inhibitor is canonical inhibitor, as it forms a tight complex with the target Proteinase which only slowly dissociate into mixture of cleaved (between its P1 and P1' residue) and uncleaved forms (Iwanaga *et al.*, 1999). Mutation studies of *Erythrina variegate* seed kunitz inhibitor, by changing P1 and surrounding residue, has shown that there is 5-fold decreases in KI (Iwanaga *et al.*, 1998).

2) Bowman Birk Inhibitors (BBI- PI) family

The Bowman Birk family is abundantly found in legumes and cereals, with reports from other plant families which have not yet been substantiated. Presence of these inhibitors have been initially detected in seeds, their induction in leaves upon wounding is also reported. The family is named after D.E. Bowman and Y. Birk, who were the first to identify and characterize a member of this family from soybean (*Glycine Max*). The soybean inhibitor is amongst the most well studied member of this family and is often referred as classic BBI. Small cyclic inhibitors have also been identified in flowers of *Helianthus annus* called sunflower trypsin inhibitor 1 (SFTI-1) (Luckett *et al.*, 1999; Zablotna *et al.*, 2002). The inhibitor acts like a substrate for the enzyme and hence called as canonical (substrate like) inhibitors. The BBIs from dicotyledonous plant typically have a molecular mass of ~ 8 kD. They are double headed and can inhibit two serine proteinases at same time. The BBI from monocotyledonous plants are of two types, one group consist of single polypeptide chain with molecular mass of 8 kD and have single reactive site. Another group has molecular mass of 16 kD with two reactive sites. Dicot 8 kD double-headed with two homologous domain, carries separate reactive site for the cognate proteases. It has been suggested that larger inhibitors were derived from smaller inhibitors by an evolutionary mechanism of gene duplication. The loss of the second domain in the small monocot inhibitor and the gene duplication event that formed the larger double domain inhibitor has been suggested to have occurred after divergence of the monocot and dicot. The first reactive site in these inhibitors is usually specific for trypsin, chymotrypsin and elastase. The active site configuration is these inhibitors are stabilized by the presence of seven conserved disulfide bond (Chen *et al.*, 1992; Lin *et al.*, 1993). In case of double headed BBIs relative affinity of binding of protease is altered of which one site is already occupied as evident by the behaviour of peanut (*Arachis hypogoea*) inhibitors. These inhibitors exhibit no activity against chymotrypsin which is pre occupied with trypsin and vice versa (Tur *et al.*, 1972). In same way the activity of soybean BBIs decreases 100-fold when trypsin in bound at other site. The BBIs family of protease inhibitors contains a unique disulphide linked nine residue loop that adopts a characteristic canonical confirmation, this loop is called protease binding loop.

3) Squash Family

This is plant unique inhibitor family; they are characterized only from the seeds of Curbitaceae family. Seven Serine PIs belonging to this family have been isolated and characterized from seed of wild cucumber (*Cyclanthera pedata*). It is smallest inhibitor family known having only 28 to 30 amino acids and P_1 at ~5th aa from N terminal. They have 3 disulphide bridges and fold in novel knottin structure (Le Nguyen *et al.*, 1990; Heitz *et al.*, 2001) with similarity to potato carboxy peptidase inhibitor. Two different but inter convertible (Cis-trans isomers) inhibitor have been isolated and characterized from seeds of wax gourd (*Benin Casa hispida [Thumb] cong*) (Atiwietin *et al.*, 2006). Because of small size of these inhibitors, studies have been carried out on native inhibitors, variants and chimera that have been prepared by chemical synthesis. The structure of squash inhibitors and inhibitor and proteinase complex has been determined by X-ray crystallography and NMR spectroscopy. These inhibitors have been shown to follow the standard mechanism for inhibition. X-Ray crystallography and NMR analysis shows that it has ellipsoidal shape; interpreted as lacking helices and β- sheets but consisting of turns and short interconnecting polypeptide structure or possessing very short helix [3 aa] and triple standard β-sheet (2,2 and 5 aa) (Bode *et al.*, 1989). Again the typical protrude reactive site loop is inserted into the proteinase active site and cleavage occur residue adjacent, with H-bond patterning virtually identical to those of all serine proteinase-protein inhibitor complex. Changes at P_1 residue that causes predictable changes in the inhibitors specificity, which are consistent with standard mechanism. Natural squash inhibitors of trypsin have arginine or lysine at P_1, where as elastase inhibitors have leucine. The P_1 arginine of several inhibitors have been altered to alanine or valine to create elastase inhibitors and to phenylalanine to make chymotrypsin inhibitors (Favel *et al.*, 1989; Min-Hua Ling *et al.*, 1993).

4) Serpin family

Widespread among the organism, being the only proteinase inhibitor found in eukarya, bacteria, arechea and viruses. Whereas prokaryotes generally have single serpin gene, large multicellular eukaryotic organism, on the other hand have several gene. Plant serpins have been purified and characterized from cereals seed, pollens and from the phloem exudates of *Cucurbita maxima* (Ostergaard *et al.*, 2000).

Serpins of Arabidopsis have been shown to act on metacaspases like protein *in vivo* and play a role in the plant immune response. It has been suggested that rather than directly interacting with pathogen, plant serpins may have role in the complex pathway involved in up regulating the host immune response (Vercammen *et al.*, 2006). Plant serpins appears to be exclusively inhibitors of serine proteinases, including trypsin, elastase and chymotrypsin. But serpins that inhibit casapases and papain like cysteine proteases have been also reported. Irreversible inhibition and the inhibitor - proteinase complex is stable to boiling in SDS which is indicative of covalent acyl bond formation. The inhibitor cleavage position is within the same structural motifs as other. They have mixed proteinase specificities like Barley Serpin (BSZx) is potent inhibitor of trypsin and chymotrypsin at overlapping reactive site (Dahl *et al.*, 1996). Plant serpins shows a typical irreversible inhibition with a second order association rate constant Ka of about $10^5 M^{-1} S^{-1}$ indicative of inhibition. The cleavage of appropriate peptide bond in reactive centre loops of inhibitor so that catalysis does not proceed beyond the formation if an acyl enzyme complex (Huntington *et al.*, 2000).

5) Ragi seed trypsin/ α-amylase inhibitor family

The cereals trypsin/ α-amylase inhibitor family comprises of single chain protein having molecular mass of about 13 kD and 5 disulfide bonds. Family members have serine protease inhibitor activity, α-amylase inhibitory activity or both. A large number of inhibitor in this family have only α-amylase inhibitor activity. However, inhibitors from barley (*Hordeum vulgare*), rye (*Secale cereals*) and tall fescue (*Festuca arundinacea*) are active against trypsin (*Elusine coracana*) inhibitor show dual activities and can inhibit serine proteases as well as α-amylase (Mahoney *et al.*, 1984). The vast majority of characterized inhibitors have dual activities. The vast majority of characterized inhibitors from this family have only α amylase inhibitory activity suggesting that serine proteinase inhibitory activity has evolved more recently. The structure of ragi inhibitor was initially solved by NMR and subsequently by X–ray analysis with the *Tenebrio molitor* α-amylase complex (Strobl *et al.*, 1995). Proteinase binding loop appeared to be adopted a canonical confirmation. Inhibitors consist of single polypeptide containing 5 disulfide bonds with a molecular mass of about 13 kD. From this structure it is also shown that proteinase binding loop is in the canonical confirmation for proteinase inhibitors, with the arginine at P_1 of the sessile bond fully exposed and projecting away from the body of the molecule. Both ragi and maize inhibitor have a novel structure containing four alpha helices and two short antiparallel β strand similar to that of the α-amylase inhibitor. This type of inhibitor has reversible inhibition mechanism with an equilibrium constant between 0.1 to 10 dependants on pH.

6) Mustard seed trypsin inhibitor family (Sinapis)

These type of inhibitors mostly found in crucifera species. These are small inhibitor with molecular weight around 7 kD. It form tight binding complex with trypsin and being cleaved by trypsin (Ceciliani *et al.*, 1994). They are wound inducible inhibitor. They show divergent evolutionary relationship with Bowman Birk Inhibitor with 27% sequence similarity. The structure has been solved using NMR, the structure has an alpha helix and two beta sheet stabilized by 4 disulfide bonds (Zhao *et al.*, 2002).

7) Potato type I PIs (Pin-I)

The inhibitors of this family are widespread in plants and have been described in many species, including potato tubers, tomato fruit, squash phloem exudates and in tomato leaves in response to wounding (Ryan, 1962). These inhibitors have the molecular mass of ~8 kD and are generally monomeric. While the inhibitors from cucurbit and potato tubers contain a single disulphide bond, the inhibitors in this family in general lack any disulphide bridges (Cai *et al.*, 1995). The inhibitory mechanism in this family is considered to fit the standard model.

X-Ray crystallographic structure of Potato type I Inhibitor protein (Pin I), both free inhibitor and inhibitor fragment also have been solved (CI-2, CMTi-V). The inhibitor has constrained loop protruding from surface of molecule which interacts directly with active site of proteinase. Comparison between free and bound inhibitor shows that there are only small conformation changes in the inhibitor upon binding, with increased stability if reactive loop (Maphalen *et al.*, 1988). Reactivity and specificity of reactive loop is varying according to P_1 residue, for example, presence of methionine or leucine at P_1 position inhibits chymotrypsin. In case of barley CI-2 inhibitor which has specificity for subtilisin shows alanine at P_1 and in eglin C with anti elastase activity containing leucine at P_1 site. The identity of these residues corresponds to the known cleavage specificities of proteinases and where changes to the P1 residue have been made (Longstaff *et al.*, 1990).

8) Potato type II PIs (Pin-II)

These protease inhibitors are expressed as precursor protein containing multiple inhibitory repeat domains (IRDs), which is further processed to release individual inhibitory unit or domain. These inhibitory repeat domain shows specificity with insect gut proteases, which signifies co-evolutionary response of plant against insect pests in due course of time (Mishra M et al, 2010). These precursor proteins get processed in the plant itself and in the insect gut upon ingestion by the action of proteases. The members of this group have been reported only from the members of Solanaceae family. Initially characterized from potato tubers, these inhibitors have been found in leaves, flowers, fruit and phloem of other solanaceaous species (Liang *et al.*, 2005). A low molecular-mass inhibitor of this family has been found to be constitutively present in Jasmme tobacco (*Nicotiana alata*) flowers (Atkinson *et al.*,

1993). Six small wound-inducible proteinase inhibitors of this family have been reported from tobacco leaves (Barrette-Ng *et al.*, 2003; 2003). An analysis of these inhibitors and genes has shown that they are composed of multiple repeat units ranges from 1 to 8 IRDs. Inhibitors of this family have been reported to inhibit chymotrypsin, trypsin, elastase, oryzin, pronase E and subtilisin.

They are best known representatives are the wound-induced proteinase inhibitors which contain up to eight sequence-repeats (the 'IP repeats') coded by the second exon of the gene. The architecture of the Pin-II genes is conserved: the first exon encoding the N-terminus of the signal peptide, and the second, major exon encoding the C-terminus of the signal peptide and a variable number of IP repeats, are always separated by a type I intron of 100–200 bp. However, the sequence of the IP repeats is variable in different proteins, presence of eight cysteine residues constituting four disulfide bridges and a single proline residue are conserved throughout the known repeat sequences.

Precursor proteins of Pin-II type inhibitors in various plants consist of 1 to 8 IRDs, which upon cleavage by endogenous proteases release single inhibitor protein active against one or multiple serine proteinases (**figure 6**) (Heath *et al.*, 1995, Tamhane et al, 2007) The ancestral Pin-II inhibitors of single inhibitory repeats are distributed in several mono- and dicotyledonous plants, whereas the multiple inhibitor repeat precursors arising from a series of gene duplication and/or domain duplication events are restricted to Solanaceae (Barta *et al.*, 2002). The Pin-II precursor PIs proteins have a conserved structure with putative endoplasmic reticulum signal peptides at the N-terminal followed by inhibitory repeat domains (IRDs) (1 to 8) and a C-terminal region, which probably functions as a vacuolar sorting signal. Pin-II type PIs have been isolated and characterized from different Solanaceous plants. NMR-based structures of series of 6KD Nicotiana sp PIs have been determined, while the structure of the precursor PIs has also been reported in *N. alata* PIs and tomato PI–II by respectively (Nielsen *et al.*, 1995; Scanlon *et al.*, 1999). A circular conformation of 43KD inhibitory precursor of *N. alata* acquired by joining the partial repeats at the N- and C-terminals generates the sixth functional IRD. This type of 'clasped bracelet' fold has been adopted by the members of Pin-II family regardless of the number of inhibitory repeat. In particular, *C. annuum* Pin-II type PI proteins and genes have been identified and characterized. A 6-KD protein PSI1.1 purified from seeds has shown the ability to inhibit trypsin and chymotrypsin (Moura and Ryan, 2001). In another study, a protein PSI1.2 from seeds has been reported to be a circularly permuted, ancestral member of Pin-II family. Seven small 6-KD PIs from leaves of *C. annuum* have been purified and the N-terminal sequenced to show homology to various repeat regions of previously reported *C. annuum* PI genes. Three repeat PI genes with the NCBI accession numbers AF221097 and AF039398 have also been isolated from cDNA libraries constructed from leaves and pericarp of *C.annuum* infected with the TMV. These CanPI genes are expressed constitutively as well as in significant amounts upon induction by various elicitors. The potential of these proteins for inhibiting insect growth and development has been demonstrated in several transgenic plants (Heath *et al.*, 1997; Sin *et al.*, 2004). Tobacco plants over-expressing potato Pin-II gene reduced the feeding and larval growth of *Manduca sexta* (Johnson *et al.*, 1989). Similarly, transgenic rice plants expressing potato Pin-II gene were protected from *Sesmia inferens* infestation, a major lepidopteran pest of rice (Duan *et al.*, 1996).

Strategical use of protease inhibitors for crop improvement in integrated pest management

Biotechnological and molecular Biology approaches have opened new avenues in the field of crop protection strategy. Genetic engineering has allowed integration of non native gene of desired trait into crop plants resulting in the enhancement of productivity. Transgenic plants have gained much attraction over a period of time for developing resistance against insect attack. Studies carried on *Bacillus thuringiensis* transgenic plants have been proved to be an effective strategy in integrated crop management. This strategy is applied to protect cotton and corn plants expressing *Cry* genes from the damage of bollworm and corn borer as evident by their low rate of survival if compared to non Bt plants. Success with Bt further led to the development of transgenic chickpea plant expressing Cry1Ac or Cry2Aa or both proteins. Though Bt technology have emerged as a potential breakthrough in crop protection management, prevalence of Bt have raised concern regarding development of Bt resistant insects (Kanglai *et al.*, 1996). Thus the deployment of integrated crop management strategies such as refugee and pyramiding of gene in transgenic plants will ensure durable insect resistance.

Protease inhibitors have attained considerable interest for its defensive role against insect pests and investigated as a viable alternative to Bt. These ubiquitous regulators of proteolytic enzymes have readily been identified as potential candidates for the development of transgenic plants. Many studies have revealed their effectivity on insect's proteases by conducting *in vitro* assays and incorporation of PI in diet. Transgenic rice, cotton, tomato, tobacco, potato, and many more have been developed and their role is assessed in response to insect attack **[Table 1]**. The first insect-resistant transgenic plants expressing a PI were produced in 1987 by expressing cowpea trypsin inhibitor (cysteine Protease inhibitor) in tobacco plant (Hilder *et al.*, 1987). Several serine PIs, the most abundant PI have also been expressed in transgenic plants for resistance against insect pests of the order Lepidoptera while cysteine PIs have been expressed against the coleopteran pests. Inhibition in growth of *H. armigera* is observed when they are allowed to feed on transgenic cotton and peas (Julia *et al.*, 1999). Transgenic rice harbouring potato protease inhibitor type II have shown resistance from pink stem borer (*Sesamia inferens*) (Duan *et al.*, 1996). Large scale field trials are currently being conducted from transgenic rice carrying serine protease inhibitor (Urte Schlulter *et al.*, 2010). However, yet no single transgenic crop plant expressing protease inhibitor gene(s) is available for farmers. This suggests we need further research and design of appropriate strategies for Pi-based insect resistance.

Gene pyramiding

Application of integrated pest management strategies is essential in order to delay insect adaptation to insecticidal protein whether it is Bt or PI. Researchers have developed transgenic plants carrying two or more insect resistant genes referred to as pyramiding of genes. The unique feature of gene pyramiding defines the long term insect control. Thus simultaneous expression of two or more resistant trait will prove an effective way to inhibit the insect adaptation. Complete resistance to diamondback moth (*Plutella xylostella*) is observed in transgenic Brocolli lines expressing Cry1A and Cry1C gene (Cao *et al.*, 2002). In an interesting study performed by *Dunse et al* transgenic cotton plants were developed expressing Pin I and Pin II family genes in combination. *N. alata* protease inhibitor (NaPI) belong to Pin II family contains four trypsin inhibitory domains and two chymotrypsin domain. They have observed the induction of resistant chymotypsin in *Helicoverpa species* upon exposure to NaPI. Thus in order to provide complete protection from *H. armigera* and *H. punctigera* Pin I type PI from *solanum tuberosum* (St-Pin1A) is used in combination with NaPI which ensured full protection (Dunse *et al.*, 2010). In the present chapter though much emphasis has been given on inhibitory role of protease inhibitor against insects, their effect of phytopathogenic microorganisms is also reported in various studies. Combined expression of sporamin (trypsin inhibitor) and cysteine PI from taro is expressed in transgenic tobacco plants exhibits inhibition of *H. armigera* and other phytopathogens *Erwinia carotovora* and *Pythium aphanidermatum*. Thus the selection of two or more different traits is essentially critical to develop an efficient resistance mechanism in plants. It is evident from above studies that gene pyramiding in transgenic plants is an effective tool to control long term insect adaptation.

Other applications as alternative to transgenic

Transgenics have been proven to be the most promising approach in recent times to develop insect resistant crops. However, existence of other alternatives is reported especially when various regulatory issues have been posed for the employment of transgenics. Most insecticides are produced as emulsifiable concentrates, suspension concentrates, wettable granules or wettable powders. Previously chemical and botanical insecticidal formulations were prepared to control lepidopteran infestation (Dadang *et al.*, 2009; Zhao *et al.*, 2009; Men *et al.*, 2008) for e.g. cyalothrin microemulsion formulation, Bt formulation, mixture of leaf extracts. Since last five decades Bt spray formulations are being used in field to control several insect pests. Bt sprays are commercially available with different brand names like Delfin, Dipel, etc. Applicability of similar formulations can be extended by using other bio based defensive molecules such as protease inhibitors. The possible applications for these systems intended to form products with an extended shelf life and easy to use for the delivery of active ingredients. Attempts to develop insecticide formulations were centred around its non toxicity to plants, maintaining active ingredient concentration and stability. These systems allow the flexibility in the use of various defensive molecules in the same system or incorporation of more than one defensive molecule for the efficient plant protection. Till now no reports on the use of protease inhibitor spray formulation is reported. Development of PI based formulation will be fruitful in order to provide resistance against plant pests.

Limitations of protease inhibitors

Upon single herbivore attack several intricate activities occurs in plants in order to protect themselves. Owing to the dynamics and complexity between plants and herbivore it is complicated to study their interaction from single outlook. Plants accumulate defensive compounds not only in herbivore-damaged leaves but also in distal intact (systemic) leaves. Clearly, a signal travels to other parts of the plants and transmits an herbivory alert. Although this systemic response was found almost forty years ago, the identity of the mobile signal remains unknown (Dieter Hermsmeier *et al.*, 2001). Thus in an attempt to generate a transgenic plants effect of the insertion of gene may bring about various physiological changes in plants via cross signalling which is needed to be considered and regulated.

Insect response to PIs

Darwin's theory of "survival of the fittest" defines the molecular, phenotypical, and physiological changes to enable an organism to compete with other species in the population in response to their changing environment. This theory is not only applied to higher organisms but lower species as well. In due course of time plants and insects have co-evolved in order to survive and to obtain maximum benefit of their habitat. Plants have developed various complicated mechanisms to protect themselves from genetical and molecular perspective of which synthesizing protease inhibitor in one such mechanism that is detailed above. Insects, however, are nowhere remains defeated, and have evolved to counter adapt plant responses. Despite the dramatically raised plant PI levels in defence against herbivorous insects, the effects on herbivore mortality or development are often relatively minor or even absent. Yet, even small effects on development may still be ecologically relevant, and affect the next generation with severely reduced fecundity (De Leo *et al.*, 2002). However, insects respond to the exposure by PIs by inducing over expression of those proteases that are of same class but normally are in less abundance or over expression of target proteases (Cloutier *et al.*, 2000). *H. armigera* when allowed to feed on giant taro PI (GtPI) decrease in total proteinase activity was measured but analysis of spectrum of digestive proteases in midgut revealed the increase in the chymotrypsin and elastase activity and decrease in the trypsin activity (Yingru Wu *et al.*, 1997). Insects have developed other strategies as well to combat the inhibitory effect or PIs such as use of presence of diet induced weakly sensitive proteases that are able to degrade PIs (Brunelle *et al.*, 2004; Gruden *et al.*, 2004; Koo *et al.*, 2008). Insects may also make use of complex proteases belong to different functional classes that will compensate for the lost activity of proteases (Kiggundu *et al.*, 2010). Small sequence variation or mutation in proteases may lead to the inefficient binding of PIs with proteases contributing PI resistance (Dunse *et al.*, 2010). Such evolutionary mechanism allows insects to adapt on previously reported as potent PIs.

Transgenic issue

Taking into account the different modes of action and activity ranges of recombinant pesticidal proteins expressed in transgenic crops, the striking complexity of biotic interactions and food web relationships in agroecosystems, and the random insertion of transgene sequences in recipient plant genomes, three main factors generally determine the environmental impact of a given pest- (or pathogen-) resistant transgenic crop: (i) the overall efficiency of the introduced resistance trait against the target herbivore (or pathogen); (ii) the activity range or functional specificity of the recombinant trait; and (iii) the (bio)chemical composition and physiological status of the host plant following transgene insertion and expression:

1) Overall efficiency of the recombinant trait

Currently, most transgenic crops grown in agricultural fields worldwide are used for weed or herbivorous insect control (James *et al.*, 2009). By definition, any pest control measure adopted in the field, whether relying on transgenic crop lines or not, may exert direct and indirect effects on microbial, animal, and plant communities. The direct effects of insect-resistant (e.g. Bt toxin-expressing) plants are limited essentially to target pest populations, but the high pesticidal efficiency of these plants may indirectly impact the fitness of non-target organisms and the overall organization of non-target populations in the field. A well-known example of this is the negative impact of Bt toxin (Cry protein)-expressing plants on arthropod parasitoids and predators provided with 'poor quality' herbivore hosts or preys suffering recombinant toxin ingestion (Naranjo *et al.*, 2009). Another example is the adjustment of secondary herbivore, auxiliary carnivore, and soil detritivore arthropod populations in agricultural fields due to the efficient repression of primary herbivore pests with Bt toxin-expressing lines (Marvier *et al.*, 2007; Cloutier *et al.*, 2008).

2) Functional specificity of the recombinant trait

Herbivore pest resistance traits introduced into crop plant genomes are usually selected for the control of a specific pest, but direct unintended effects on non-target organisms cannot be excluded (Groot *et al.*, 2002). The monarch butterfly controversy and the reported toxicity of Cry1A and Cry2A toxins against different lepidopteran insects are examples of the preferential, but non-exclusive action of pesticidal proteins against specific target pests (Sims *et al.*, 1995; 1996; Van Frankenhuyzen *et al.*, 2002).

3) Alteration of the host plant's characteristics

The heterologous expression of metabolic regulators such as protease inhibitors in transgenic crops also raises questions about their effects in *planta* and their resulting impact on the composition and physiology of the host plant, which represents a central element in the whole food chain (Visal-Shah *et al.*, 2000; Khalf M *et al.*, 2010). Unlike Cry toxins acting on specific membrane protein receptors in the digestive tract of target herbivores protease inhibitors may interact with plant endogenous protease targets structurally and functionally related to herbivore digestive proteases (Pigott *et al.*, 2007; Goulet *et al.*, 2008). Plants possess diverse type of proteases which functions endogenously in plant metabolism and growth. Thus transgenic plants harbouring proteinase inhibitors of different functional classes though will be efficient in inhibiting insects, their effect on plant native proteases is of major concern. Further, it is also essential to assess their effect on humans and other organisms.

Future perspective

Employment of protease inhibitors highlights an effective strategy in insect crop management owing to its potential impact on agronomics. Several studies have reported the potency of protease inhibitors in retarding growth and development of insects but only few studies have demonstrated the generation of novel inhibitors by protein engineering for enhancing specificity and potency against mixture of gut proteases. Diverse type of isoforms or variants of protease inhibitors have been reported which display sequence and functional differences that ultimately leads to the assorted interaction between the two. Selection of variants that exhibits increased potency against target proteases and less effectivity against non target protease can be done by identifying relevant target sites for site directed mutagenesis by performing rational designing based on 3-D models of protease inhibitors: proteases complexes (Duan *et al.*, 1996). Characterization of the molecular basis of resistance of an insect trypsin/ chymotrypsin to the plant proteinase inhibitor is required (Dunse *et al.*, 2010). Understanding how insects develop resistance to PIs is fundamental to the future rational design of PIs for the protection of agronomically important plants against insect pests. This knowledge will allow the development of a strategy to produce lepidopteran resistant transgenic plants that will incorporate multiple inhibitors with structural diversity that target different enzyme specificities. Constructing synthetic

PIs with high specificity will also be beneficial in terms of delaying resistance by insects. *In vitro* molecular evolution schemes involving recombination or random mutagenesis in functionally relevant regions of the gene (protein) sequence, combined with high throughput screening approaches such as molecular phage display for the selection of improved inhibitor variants, probably represent an effective way to generate functionally diverse inhibitor populations. Insects response to plant PIs by inducing insensitive or different functional class of proteases can be understand by the knowledge of insect responsive promoters which will not only allow researchers to engineer insect-resistance, but also to analyze the contribution of individual transgenes to the darwinian fitness of natural plant populations exposed to herbivory.

The Council of Scientific and Industrial Research (CSIR), Government of India, New Delhi supported this project under network project grants to National Chemical Laboratory, Pune (NWP0003). Neha Khandelwal and Rakesh S. Joshi acknowledge CSIR for the research fellowships.

References:

Dileep K. singh, Applied entomology, Insecticidal methods for pest control

Reddy, K.V.S. and Zehr U.B. (2004) Novel strategies for overcoming pests and diseases in India. *Proceedings of the 4th International Crop Science Congress.*

Rao, R.G.V., Rao, R.V. and Nigam, S.N. (2010) Post-harvest Insect Pests of Groundnut and their Management. *International crop research institute for the semi arid tropics.*

Mayuri, K. and Mikunthan, G. (2009) Damage pattern of Cabbage flea beetle, *Phyllotreta cruciferae* (Goeze) (coleoptera: chrysomelidae) and its associated hosts of crops and weeds. *American-Eurasian J. Agric. & Environ. sci.*, **6 (3)**, 303-307.

Sharma, M., Charak K.S. and Ramanaiah T.V. (2003) Agricultural biotechnology research in India: Status and policies. *Current science*, **84 (3)**, 297-302.

Hanur, V.S. (2008) Bt resistance and monophagous pests: Handling with prudence. *Current Science.* 95 (4).

Host plant or insect- Survival of the fittest (anonymous)

Showler, A.T. (2001) *Spodoptera exigua* ovisposition and larval feeding preferences for pigweed, *Amaranthus hybridus*, over squaring cotton, *Gossypium hirsutum*, and a comparison of free amino acids in each host plant. *Journal of Chemical Ecology*, **27 (10)**, 2013-28

Tilmann, P.G. (2006) Tobacco as a trap crop for *Heliothis virescens*. *J. Entomol. Sci.*, **41(4)**, 305-320.

Bennett, R.N. and Wallsgrove, R.M. (1994) Secondary metabolites in plant defense- mechanisms. *New Phytologist*, **127**, 617–633.

Kessler, A. and Baldwin, I.T. (2001) Defensive function of herbivore induced plant volatile emissions in nature. *Science*, **291**, 2141-2144.

Kessler, A. and Baldwin, I.T. (2002) Plant responses to insect herbivory: the emerging molecular analysis. *Annual Review in Plant Biology*, **53**, 299-328.

Ferry, N., Edwards, M.G., Gatehouse, J.A. and Gatehouse, A.M. (2004) Plant-insect interactions: molecular approaches to insect resistance Current Opinion in Biotechnology. *Curr Opin Biotechnol.*, **15**, 155-161.

Kang, J.H., Wang, L., Giri, A.P. and Baldwin, I.T. (2006) Silencing threonine deaminase and JAR4 in *Nicotiana attenuata* impairs jasmonic acid-isoleucine-mediated defenses against *Manduca sexta*. *The Plant Cell*, **18**, 3303-3320.

Orozco-Cardenas, M. and Ryan, C.A. (1999) Hydrogen peroxide is generated systemically in plant leaves by wounding and systemin via the octadecanoid pathway. *Proceedings of the National Academy of Sciences of the United States of America*, **96**, 6553-6557.

Orozco-Cardenas, M.L., Narvaez-Vasquez, J. and Ryan, C.A. (2001) Hydrogen peroxide acts as a second messenger for the induction of defense genes in tomato plants in response to wounding, systemin, and methyl jasmonate. *The Plant Cell*, **13**, 179-191.

Rojo, E., Solano, R. and Sanchez-Serrano, J.J. (2003) Interactions between signaling compounds involved in plant defence. *J Plant Growth Regul.*, **22**, 82-98

Boulter, D., Gatehouse, J.A., Gatehouse, A.M.R. and Hilder, V.A. (1990) Genetic engineering of plants for insect resistance. *Endeavour*, **14(4)**, 185-190.

Jongsma, M.A. and Beekwilder, J. (2008) Plant Protease Inhibitors: Functional Evolution for Defense. *Springer Netherlands*, 235-251.

Tamura, T., Hara, K., Yamaguchi, Y., Koizumi, N. and Sano, H. (2003) Osmotic Stress Tolerance of Transgenic Tobacco Expressing a Gene Encoding a

Membrane-Located Receptor-Like Protein from Tobacco Plants. *Plant Physiol.*, **131(2)**, 454-462.

Markus, H., Giri, A.P., Kaur, H. and Baldwin, I.T. (2010) Serine Protease Inhibitors Specifically Defend *Solanum nigrum* against Generalist Herbivores but Do Not Influence Plant Growth and Development. *The Plant Cell*, **22**, 4158–4175

Broadway, R.M. and Duffey, S.S. (1986) Plant proteinase inhibitors: Mechanism of action and effect on the growth and digestive physiology of larval *Heliothis zea* and *Spodoptera exigua. J. Insect Physiol.*, **32**, 827-833.

Heath, R.L., McDonald, G., Christeller, J.T., Lee, M., Bateman, K., West, J., Heeswijck, R.H. and Anderson, M.A. (1997) Proteinase inhibitors from *Nicotiana alata* enhance plant resistance to insect pests. *Journal of Insect Physiology*, **43(9)**, 833-842.

Fan, S.G., and WU, G.J. (2005) Characteristics of plant proteinase inhibitors and their applications in combating phytophagous insects. *Bot. Bull. Acad. Sin.*, **46**, 273-292.

Lawrence, P.K. (2002) Plant protease inhibitors in control of phytophagous insects. *Electronic Journal of Biotechnology*, **5 (1)**.

Arai, S., Watanabe, H., Kondo, H., Emori, Y., and Abe, K. (1991) Papain inhibitory activity of oryzacystatin, a rice seed cysteine proteinase inhibitor, depends on the central Gln-Val-Val-Ala-Gly region conserved among cystatin superfamily members. *Biochem.*, **109** (2), 294-298.

Huma, H. and Fazili, K.M. (2007) Plant protease inhibitors: a defense strategy in plants. *Biotechnology and Molecular Biology Review*, **2(3)**, 068-085.

Graham, J.S. and Ryan, C.A. (1997) Accumulation of metallocarboxy -peptidase inhibitor in leaves of wounded potato plants. *Biochemical and Biophysics Research Communication*, **101**, 1164-1170.

Hollander-Czytko, H. Andersen, J.L. and Ryan, C.A. (1985) Vacuolar localization of wound-induced carboxy peptidase inhibitor in potato leaves. *Plant Physiology*, **78**, 76-79.

Liang, W.A., MacManus, M.T. and Allan, A.C. (2002) Protein-protein interaction in plant. *Annual Plant Reviews*, **7**, 77-119.

Ritonja, A., Krizaj, I., MeSko, P., Kopitar, M., Lucovnik, P., Strukelj, B., Pungercar, J., Buttle, D. J., Barrett, A. J. and Turk, V. (1990) The amino acid sequence of a novel inhibitor of cathepsin D from potato. *FEBS Letters*, **267**, 13-15.

Brzin, J., Popovi, T., Drobnic-Kosorok, M., Kotnik, M. and Turk, V. (1988) Inhibitors of cysteine proteinases from potato. *Biological Chemistry Hoppe-Seyler*, **369**, 233-238.

Song, H.K. and Suh, S.W. (1998) Kunitz-type soybean trypsin inhibitor revisited: refined structure of its complex with porcine trypsin reveals an insight into the interaction between a homologous inhibitor from Erythrina caffra and tissue-type plasminogen activator1. *J. Mol Bio.*, **275(2)**, 347-363.

Iwanaga, S., Nagata, R., Miyamoto, A., Kouzuma, Y., Yamasaki, N. and Kimura, M. (1999) Conformation of the primary binding loop folded through an intramolecular interaction contributes to the strong chymotrypsin inhibitory activity of the chymotrypsin inhibitor from Erythrina variegata seeds. *J Biochem*, **126(1)**, 162-7.

Iwanaga, S., Yamasaki, N. and Kimura, M. (1998) Chymotrypsin inhibitor from Erythrina variegata seeds: involvement of amino acid residues within the primary binding loop in potent inhibitory activity toward chymotrypsin. *J Biochem.*, **124(3)**, 663-9.

Luckett, S., Garcia, R.S., Barker, J.J., Konare, A.V., Shewry, P.R., Clarke, A.R. and Brady, R.L. (1999) High resolution structure of a potent, cyclic proteinase inhibitor from sunflower seeds. *J Mol. Bio.*, **290**, 525-533.

Zablotna, E., Kazmierczak, K., Jaskiewicz A., Stawikowski, M., Kupryszewski, G., and Rolka, K. (2002) Chemical synthesis and kinetic study of the smallest naturally occurring trypsin inhibitor SFTI-1 isolated from sun-flower seeds and its analogues. *Biochem Biophys Res Commun.*, **292**, 855–859.

Chen, P., Rose, J., Love, R., Wei, C.H. and Wang, B.C. (1992) Reactive sites of an anticarcinogenic Bowman-Birk proteinase inhibitor are similar to other trypsin inhibitors. *J. Biol. Chem.*, **267**, 1990-1994.

Lin, G., Bode, W., Huber, R., Chi, C. and Engh, R.A. (1993) The 0.25nm X-ray structure of the Bowman-Birk type inhibitor from mung bean in ternary complex with porcine trypsin. *Eur. J. Biochem.*, **212**, 549-555.

Tur-sinal, A., Birk, Y., Gertler, A. and Rigbi, M. (1972) Basic trypsin and chymotrypsin inhibitor from groundnut (arachis hypogea). *biochim. Biophys. Acta.*, **263**, 666-672.

Nguyen, D.L., Heitz, A., Chiche, L., Castro, B., Boigegrain, R.A., Favel, A. and Coletti-Previero, M.A. (1990) Molecular recognition between serine proteases and new bioactive microproteins with a knotted structure. *Biochimie*, **72(6-7)**, 431-435.

Heitz, A., Hernadez J.F., Gagnon, J., Hong, T.T., Pham, T.T., Nguyen, T.M., Le-Nguyen, D. and Chiche L. (2001) Solution Structure of the Squash Trypsin Inhibitor MCoTI-II. A New Family for Cyclic Knottins. *Biochemistry*, **40(27)**, 7973-7983.

Atiwetin, P., Harada, S. and Kamei, K. (2006) Serine protease inhibitor from Wax Gourd (Benincasa hispida) seeds. *Biosci. Biotech. Biochem.*, **70(3)**, 743-745.

Bode, W., Greyling, H.J., Huber, R., Otlewski, J., and Wilusz T. (1989) The refined 2.0 A X-ray crystal structure of the complex formed between bovine beta-trypsin and CMTI-I, a trypsin inhibitor from squash seeds (Cucurbita maxima). Topological similarity of the squash seed inhibitors with the carboxypeptidase A inhibitor from potatoes. *FEBS Letters*, **242(2)**, 285 - 292.

Favel, A., Nguyen, D.L., Coletti-Previero, M.A. and Castro, B. (1989) Active site chemical mutagenesis of Ecballium elaterium Trypsin Inhibitor II: New microproteins inhibiting elastase and chymotrypsin. *Biochem. Biophys. Res Communication*, **162(1)**, 79-82.

Ling, M.H., Qi, H.Y., and Chi, C.W. (1993) Protein, cDNA and genomic DNA sequences of the towel gourd Trypsin inhibitor, A squash family inhibitor, *J. Biol. Chem.*, **268**, 810-814.

Østergaard, H., *et al.* (2000) Inhibitory Serpins from Wheat Grain with R active Centers Resembling Glutamine-rich Repeats of Prolamin Storage Proteins. *Journal of Biological Chemistry*, **275(43)**, 33272-33279.

Vercammen D., Beatrice, B., van de Cotte, B., Beunens, T., Gavigan, J.A., De Rycke R.D., Brackenier, A., Inze, D., Harris, J.L. and Breusegem F.V. (2006) Serpin1 of Arabidopsis thaliana is a suicide inhibitor for metacaspase 9. *J. Mol. Biol.*, **364(4)**, 625-636.

Dahl, S.W., Rasmussen, S. K. and Hejgaard, J. (1996) Heterologous Expression of Three Plant Serpins with Distinct Inhibitory Specificities. *Journal of Biological Chemistry*, **271(41)**: p. 25083-25088.

Huntington, J.A., Read, R.J. and Carrell, R.W. (2000) Structure of serpin protease complex shows inhibition by deformation. *Nature*, **407(6806)**: p. 923-926.

Mahoney, W.C., Hermodson, M.A., Jones, B., Powers, D.D. and Corfman, R.S., Reeck, G.R. (1984) Amino acid sequence and secondary structural analysis of the corn inhibitor of trypsin and activated Hageman Factor. *Journal of Biological Chemistry*, **259(13)**, 8412-8416.

Strobl, S., Muhlhahn P., Bernstein, R., Wiltscheck, R., Maskos, K., Wunderlich, M., Huber, R., Glockshuber, R. and Holak, T.A. (1995) Determination of the three-dimensional structure of the bifunctional alpha-amylase/trypsin inhibitor from ragi seeds by NMR spectroscopy. *Biochemistry*, **34(26)**, 8281-8293.

Ceciliani F, Bortolotti, F., Menegatti, E., Ronchi S, Ascenzi, P. and Pakmieri, S. (1994) Purification, inhibitory properties, amino acid sequence and identification of the reactive site of a new serine proteinase inhibitor from oil-rape (Brassica napus) seed. *FEBS Lett.*, **342**, 221-224.

Zhao, Q., Chae, Y.K., and Markley, J.L. (2002) NMR Solution Structure of ATTp, an Arabidopsis thaliana Trypsin Inhibitor. *Biochemistry*, **41(41)**, 12284-12296.

Ryan, C.A. and Balls, A.K. (1962) An inhibitor of chymotrypsin from Solanum tuberosm and its behavior toward trypsin. *Proc. Natl. Acad. Sci.*, **(48)**, 1839-44.

Cai, M., Gong, Y., Kao, J., Krishnamoorthi, R. (1995) Three-dimensional solution structure of Cucurbita maxima trypsin inhibitor-V determined by NMR spectroscopy. *Biochemistry*, **34(15)**, 5201-11.

McPhalen, C.A. and James, M.N.G. (1988) Structural composition of two serine proteinase-protein inhibitor complexes: eglin-c- subtilisin Carlsberg and CI-2 subtilisin Novo. *Biochemistry*, **27(17)**, 6582-6598.

Longstaff, C., Campbell, A.F. and Fersht, A.R. (1990) Recombinant chymotrypsin inhibitor 2: expression, kinetic analysis of inhibition with .alpha.-chymotrypsin and wild-type and mutant subtilisin BPN', and protein engineering to investigate inhibitory specificity and mechanism. *Biochemistry*, **29(31)**, 7339-7347.

Christeller, J. and Liang, W. (2005) Plant serine protease inhibitors. *Protein and Peptide Letters*, **12**, 439-447.

Atkinson, A.H., Heath, R.L., Simpson, R.J., Clarke, A.E. and Anderson, M.A. (1993) Proteinase inhibitors in *Nicotiana alata* stigmas are derived from a precursor protein which is processed into five homologous inhibitors. *Plant Cell*, **5**, 203-213.

Barrette-Ng, I.H., Ng, K.K., Cherney, M.M. and Pearce, G. (2003) Structural basis of inhibition revealed by a 1:2 complex of the two-headed tomato inhibitor-II and subtilisin Carlsberg. *J. Biol. Chem.*, **278**, 24062-24071.

Heath, R.L., Barton, P.A., Simpson, R.J., Reid, G.E., Lim, G. and Anderson, M.A. (1995) Characterization of the protease processing sites in a multidomain proteinase inhibitor precursor from *Nicotiana alata*. *Eur. J. Biochem.*, **230**, 250-257.

Barta, E., Pintar, A. and Pongor, S. (2002) Repeats with variations: accelerated evolution of the Pin2 family of proteinase inhibitors. *Trends Genet.*, **18**, 600-603.

Nielsen, K.J., Heath, R.L., Anderson, M.A. and Craik, D.J. (1995) Structures of a series of 6-kDa trypsin inhibitors isolated from the stigma of *Nicotiana alata*. *Biochemistry*, **34**: p. 14304-14311.

Scanlon, M.J., Lee, M.C., Anderson, M.A. and Craik, D.J. (1999) Structure of a putative ancestral protein encoded by a single sequence repeat from a multidomain proteinase inhibitor gene from *Nicotiana alata*. *Structure Fold. Des.*, **7**, 793-802.

Moura, D.S. and Ryan, C.A. (2001) Wound-inducible proteinase inhibitors in pepper. Differential regulation upon wounding, systemin and methyl jasmonate. *Plant Physiol.*, **126**, 289-298.

Heath R.L., McDonald, G., Christeller, J.T., Lee, M., Bateman, K., West, J., Heeswijck, R.V. and Anderso, M.A. (1997) Proteinase inhibitors from *Nicotiana alata* enhance plant resistance to insect pests. *Journal of Insect Physiology*, **43(9)**, 833-842.

Sin, S.F. and Chye, M.L. (2004) Expression of proteinase inhibitor II proteins during floral development in *Solanum americanum*. *Planta*, **219**, 1010-1022.

Johnson, R., Narvaez, J., An, G. and Ryan, C.A. (1989) Expression of proteinase inhibitors I and II in transgenic plants: effects on natural defense against *Manduca sexta* larvae. *Proc. Natl. Acad. Sci.*,USA, **86**, 9871-9875.

Duan, X., Li, X., Xue, Q., Abo-EI-Saad, M., Xu, D. and Wu, R. (1996) Transgenic rice plants harboring an introduced potato proteinase inhibitor II gene are insect resistant. *Nature Biotechnology*, **14**, 494-498.

He, K., Wang, Z., Bai, S., Zheng, L., Wang, Y. and Cui, H. (2006) Efficacy of transgenic Bt cotton for resistance to the Asian corn borer (Lepidoptera: Crambidae). *Crop Protection*, **25(2)**, 167-173.

Hilder, V.A., Gatehouse, A.M.R., sheerman, S.E., Barker, R.F. and Boulder, D. (1987) A novel mechanism of insect resistance engineered into tobacco. *Nature*, **300**, 160-163.

Charity, J.A., Anderson, M.A., Bittisnich, D.J., Whitecross, M. and Higgins, T.J.V. (1999) Transgenic tobacco and peas expressing a proteinase inhibitor from *Nicotiana alata* have increased insect resistance. *Molecular breeding*, **5(4)**, 357–365.

Schluter, U., Benchabane, M., Munger, A., Kiggundu, A., Vorster, J., Goulet, M. C., Cloutier, C., and Michaud, D. (2010) Recombinant protease inhibitors for herbivore pest control: a multitrophic perspective. *Journal of experimental botany*, **61(15)**, 4169-4183.

Cao, J., Zhao, J.Z., Tang, J., Shelton, A. and Earle, E. (2002) Broccoli plants with pyramided cry1Ac and cry1C Bt genes control diamondback moths resistant to Cry1A and Cry1C proteins. *Theor Appl Genet*, **105**, 258-264.

Dunse, K.M., Kaas, Q., Guarino, R.F., Barton, P.A., Craik, D.J. and Anderson, M.A. (2010) Molecular basis for the resistance of an insect chymotrypsin to a potato type II proteinase inhibitor. *PNAS*, **107(34)**, 15016-15021.

Dadang, Fitriasari, E.D. and Prijono, D. (2009) Effectiveness of two botanical insecticide formulations to two major cabbage insect pests on field application. *J. ISSAAS*, **15(1)**, 42-51.

Zhao, F., Xia, H.-Y. and He, J.-L. (2009) Formulation design of cyhalothrin pesticide microemulsion. *Current Science*, **97(10)**.

Mahesh, P. and Men, U.B. (2008) Effect of Bt formulations on Yield of Brinjal by Managing Leucinodes orbonalis. *Annals of Plant Protection Sciences*, **16(2)**, 485-547.

Dieter Hermsmeier, U.S., and Baldwin, I.T. (2001) Molecular Interactions between the Specialist Herbivore *Manduca sexta* (Lepidoptera, Sphingidae) and Its Natural Host *Nicotiana attenuata*. I. Large-Scale Changes in the Accumulation of Growth- and Defense-Related Plant mRNAs. *Plant physiol.*, **125(2)**, 683-700.

De Leo, F. and Gallerani, R. (2002) The mustard trypsin inhibitor 2 affects the fertility of *Spodoptera littoralis* larvae fed on transgenic plants. *Insect Biochemistry and Molecular Biology*, **32**, 489-496.

Cloutier C., Jean, C., Fournier, M., Yelle, S. and Michaud, D. (2000) Adult Colorado potato beetles, *Leptinotarsa decemlineata* compensate for nutritional stress on oryzacystatin I-transgenic potato plants by hypertrophic behaviour and over-production of insensitive proteinases. *Archives of Insect Physiology and Biochemistry*, **44**, 69-81.

Wu, Y., Llewellyn, D., Mathews, A., and Dennis, E.S. (1997) Adaptation of *Helicoverpa armigera* (Lepidoptera: Noctuidae) to a proteinase inhibitor expressed in transgenic tobacco. *Molecular breeding*, **3**, 371-380.

Brunelle F, Cloutier, C. and Michaud D. (2004) Colorado potato beetles compensate for tomato cathepsin D inhibitor expressed in transgenic potato. *Archives of Insect Physiology and Biochemistry*, **55**, 103 113.

Gruden K, Kuipers, A.G., Guncar, G., Slapar, N., Strukelj, B., Jongsma, M.A. (2004) Molecular basis of Colorado potato beetle adaptation to potato plant defence at the level of digestive cysteine proteinases. *Insect Biochemistry and Molecular Biology*, **34**, 365-375.

Koo, Y.D., Ahn, J.E., Salzman, R.A., Moon, J., Chi, Y.H., Yun, D.J., Lee, S.Y., Koiwa, H., Zhu-Salzman, K. (2008) Functional expression of an insect cathepsin B-like counter-defence protein. *Insect Molecular Biology*, **17**, 235-245.

Kiggundu, A., Muchwezi, J., Van der Vyver, C., Viljoen, A., Vorster, J., Schlü¨ter, U., Kunert, K. and Michaud, D. (2010) Deleterious effects of plant cystatins against the banana weevil *Cosmopolites sordidus*. *Archives of Insect Biochemistry and Physiology*, **73**, 87-105.

Dunse, K.M., Kaas, Q., Guarino, R.F., Barton, P.A., Craik, D.J., and. Anderson, M.A. (2010) Molecular basis for the resistance of an insect chymotrypsin to a potato type II proteinase inhibitor. *PNAS*, **107(34)**, 15016-15021.

James, C. (2009) Global status of commercialized biotech/GM crops: 2009. Ithaca, *NY: ISAAA*, ISAAA Brief No. 41.

Naranjo, S.E. (2009) Impacts of Bt crops on non-target invertebrates and insecticide use patterns. *CAB Reviews: Perspectives in Agriculture, Veterinary Science, Nutrition and Natural Resources*, 4(23).

Marvier, M., McCreedy, C., Regetz, J. and Kareiva, P. (2007) A meta-analysis of effects of Bt cotton and maize on non-target invertebrates. *Science*, **316**, 1475-1477.

Cloutier, C., Boudreault, S. and Michaud, D. (2008) Impact de pommes de terre re´ sistantes au doryphore sur les arthropodes non vise´ s: une me´ ta-analyse des facteurs possiblement en cause dans l'e´ chec d'une plante transgenique Bt. *Cahiers Agriculture*, **17**, 388-394

Groot, A.T. and Dicke, M. (2002) Insect-resistant transgenic plants in a multi-trophic context. *The Plant Journal*, **31**, 387-406.

Sims, S.R. (1995) *Bacillus thuringiensis* var. kurstaki [CryIA(c)] protein expressed in transgenic cotton: effects on beneficial and other nontarget insects. Southwestern Entomologist, **20**, 493-500.

Sims, S.R. (1997) Host activity spectrum of the CryIIA *Bacillus thuringiensis* subsp. kurstaki protein: effect on Lepidoptera, Diptera, and non-target arthropods. *Southwestern Entomologist*, **22**, 395-404.

Van Frankenhuyzen, K., Nystrom, C. (2002) The Bacillus thuringiensis toxin specificity database. Internet webpage: http:// www.glfc.cfs.nrcan.gc.ca/bacillus (accessed Ferbruary 28, 2010).

Visal-Shah, S., Brunelle, F. and Michaud, D., Multiple protease/ inhibitor interactions in plant-pest systems. *Recombinant protease inhibitors in plants, Georgetown, TX: Landes Bioscience*, 107-113.

Khalf, M., Goulet, C., Vorster, J., Brunelle, F., Anguenot, R., Fliss, I. and Michaud, D. (2010) Tubers from potato lines expressing a tomato Kunitz protease inhibitor are substantially equivalent to parental and transgenic controls. *Plant Biotechnology Journal*, **8**,155-169.

Pigott, C.R. and Ellar, D.J. (2007) Role of receptors in Bacillus thuringiensis crystal toxin activity. *Microbiology and Molecular Biology Reviews*, **71**, 255281.

Goulet, M.C., Dallaire, C., Vaillancourt, L.P., Khalf, M., Badri, M.A., Preradov, A., Duceppe, M.O., Goulet, C., Cloutier, C. and Michaud, D. (2008) Tailoring the specificity of a plant cystatin toward herbivorous insect digestive cysteine proteases by single mutations at positively selected amino acid sites. *Plant Physiology*, **146**, 1010-1019.

Patankar, A.G., Giri, A.P. Harsulkar, A.M., Sainani, M.N., Deshpande, V.V. Ranjekar, P. K. and Gupta, V.S. (2001) Complexity in specificities and expression of *Helicoverpa armigera* gut proteinases explains polyphagous nature of the insect pest. *Insect Biochemistry and Molecular Biology*, **31**, 453–464

Tamhane, V.A., Giri, A.P. Kumar, P. and Gupta, V.S. (2009), Spatial and temporal expression patterns of diverse Pin-II proteinase inhibitor genes in Capsicum annuum Linn. *Gene*, **442**, 88–98

Mishra, M., Tamhane, V.A., Khandelwal, N., Kulkarni, M.J., Gupta, V.S. and Giri, A.P. (2010), Interaction of recombinant CanPIs with *Helicoverpa armigera* gut proteases reveals their processing patterns, stability and efficiency. *Proteomics*, *10*, 2845–2857

Tamhane, V.A., Giri, A.P., Sainani, M.N. and Gupta, V.S. (2007)
Diverse forms of Pin-II family proteinase inhibitors from Capsicum annuum adversely affect the growth and development of *Helicoverpa armigera*.
Gene, 403, 29–38

Figure Legends:

Figure 1: The figure represents productivity loss of individual crop by insect pests in terms of percentage. Cotton (*) is highly affected by insect infestation leading to 50% loss of crop whereas wheat crop is least affected with total loss of 5%

Figure 2: Lifecycle of an insect includes four stages: from egg, larva and pupa to adult. Larva of the insect passes through various instars (six in lepidopteran and three in coleopteran) which allows it to attain complete maturity. Followed by pupation period adult (moth or butterfly in lepidoptera and beetles in coleoptera) emerges. Female adult lays hundreds of eggs on the leaves which further hatch into the1[st] instar larvae.

Figure 3: Plant's response to insect attack is categorized into direct and indirect defense which involves in preventing insect proliferation. Direct response comprises defensive chemicals which directly affect target pest whereas indirect defense involves attraction of herbivore predators by emission of volatiles.

Figure 4: Reflexes produced by a plant during insect infestation, which involves production of Jasmonic acid (JA) by systemin through octadecanoid pathway. JA in the form of Jasmonic Isoleucine (JA-Ile) triggers expression of early and late defense genes

Figure 5: PI exists as the plant's natural defense mechanism which upon ingestion by insects interacts with the midgut proteases. PI acts by blocking active site of protease rendering it inactive for protein catabolism. Thus, accumulation of the protein in the gut creates a physiological burden on insect ultimately leading to death of the insect pests.

Figure 6: Proteinase inhibitors are classified into four classes depending upon the proteases they inhibit. Serine protease inhibitors are further divided into various families based on their abundance, structure and function.

Figure 7: This figure represents the processing of multi inhibitory repeat domains of PI protein (IRDs) to active single inhibitory domain. **[a]** A representative PI with 4 IRD (Precursor form) upon cleavage at the linker region gets converted into single IRD (Mature form).**[b]** Mass spectrometric analysis describes the eventual release of single IRD as shown by its increased peak intensity whereas the diminishing peak intensities of higher IRD (4IRD, 3IRD and 2 IRD) forms of a specific PinII type PI, CanPI-7 from *Capsicum annum*

Table 1

Transgenics	Protease Inhibitors	Class	Target insect Pest	References
Tobacco	NA-PI	Serine	*Spodoptera litura*	Shrinivasan et al, 2009
			H. armigera	Charity et al, 1999
Peas	Na-PI	Serine	*H. armigera*	Charity et al, 1999
Oilseed rape	MTI-2	Serine	*P. xylostella*	De Leo et al., 2001
Potato	Tomato- Cathepsin d inhibitor	Cystatin	*Leptinotarsa decemlineata*	Brunelle et al, , 2004
Barley	cystatin HvCPI-1 C68	Cysteine	*Leptinotarsa decemlineata*	Alvarez-Alfageme et al., 2007
Rice	Barley TI	Serine	*Sitophilus oryzae*	Alfonso-Rubi et al., 2003
	Soybean TI	Serine	*Nilaparvata lugens*	Li et al., 2005
			Cnaphalocrocis medinalis	Li et al., 2005
	M-PI	Serine	*Chilo suppresssalis*	Vila et al., 2005
	Cowpea-TI	Serine	*C. medinalis*	Han et al., 2007
	Brassica juncea TI	Serine	*Spodoptera litura*	Mandal et al., 2002
	Mustard TI-2	Serine	*Spodoptera littoralis*	De Leo and Gallerani, 2002
	Tobacco TI	Serine	*S. litura*	Srinivasan et al., 2009
			H. armigera	Srinivasan et al., 2009
	Potato Pin II		*Sesamia inferens*	Duan X et al, 1996
	Sporamin+Taro cystatin	Serine+Cysteine	*H. armigera*	Senthilkumar et al., 2010
	Buckwheat serine PI	Serine	*Trialeurodes vaporariorum*	Khadeeva et al., 2009
Tomato	Taro cystatin	Cystatin	*M. incognita*	Chan et al 2010
Cotton	Na-PI and St Pin1A	Serine	*H. armigera*	Dunse et al, 2010

Table 1: Few of the transgenics harbouring protease inhibitor genes for the control of target pest PI transgenics

Figure 1
Wheat 5%
Rice 25%
Maize 25%
*Cotton 50%
Other cereals 30%
Chickpea 10%
Other pulses 20%
Sugarcane 20%
Groundnut 15%
Other Oilseeds 20%
Rapeseed & mustered 30%

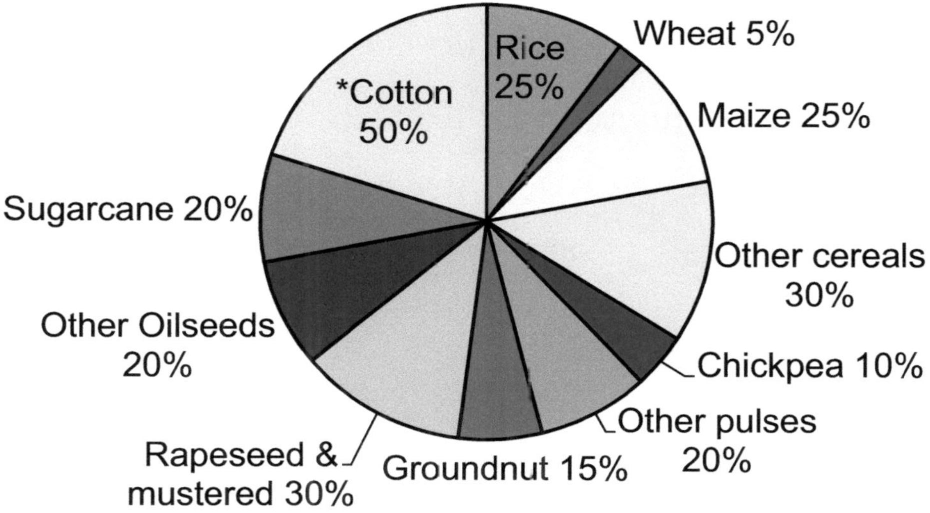

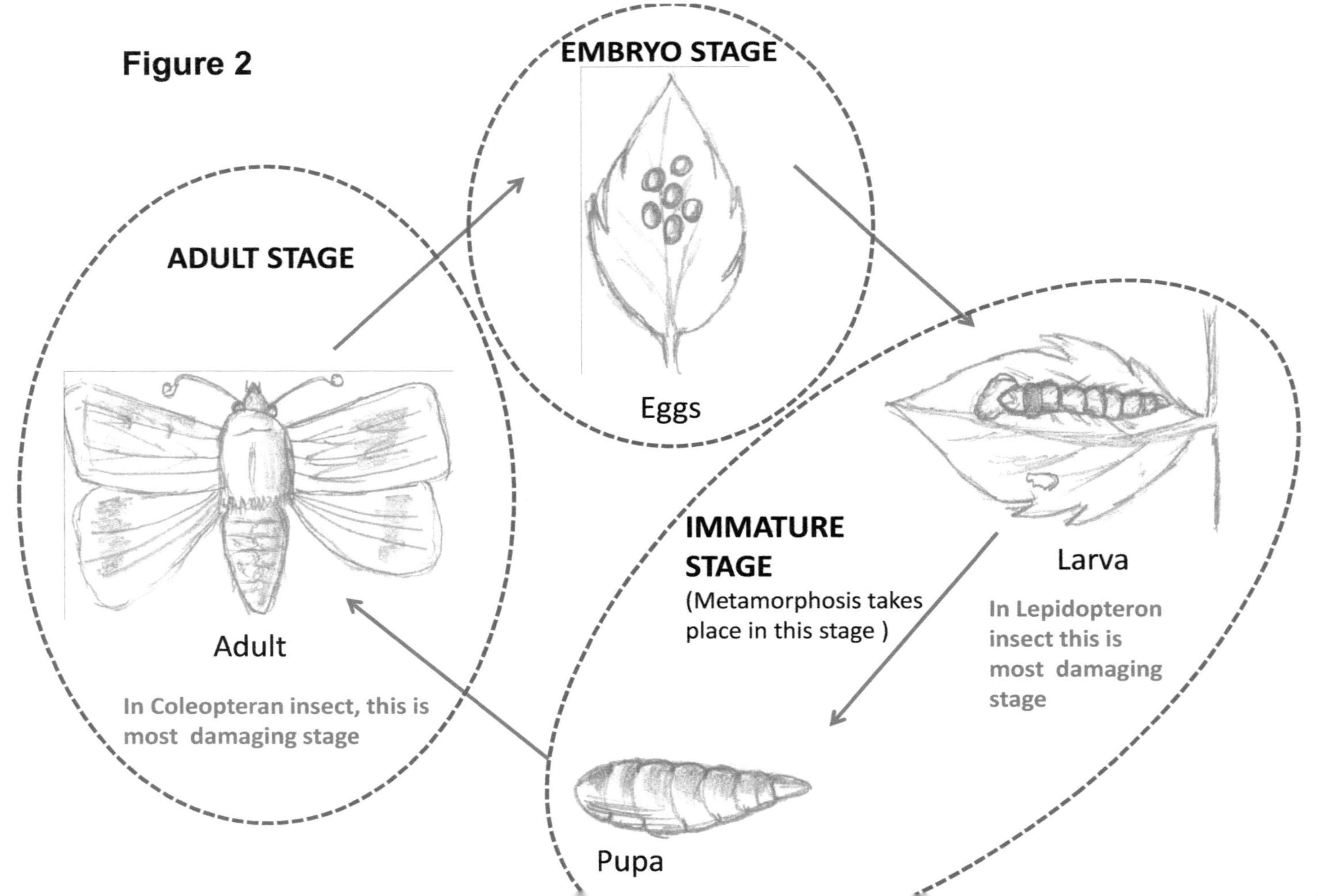

Figure 2
EMBRYO STAGE
ADULT STAGE
Eggs
Adult
In Coleopteran insect, this is most damaging stage
IMMATURE STAGE
(Metamorphosis takes place in this stage)
Larva
In Lepidopteron insect this is most damaging stage
Pupa

Figure 3

Plant Defence
Direct Defence
Indirect Defence
Physical Barrier
Toxic, Repellent or Anti-Digestive Molecules
Emit Volatiles which produces the extra floral nectar that attract predators of insect herbivorous
Cuticle
Trichomes
Thorns
Cynogenic Glucoside
Glucosinolates
Alkaloids
Phenolics
Proteinase Inhibitor

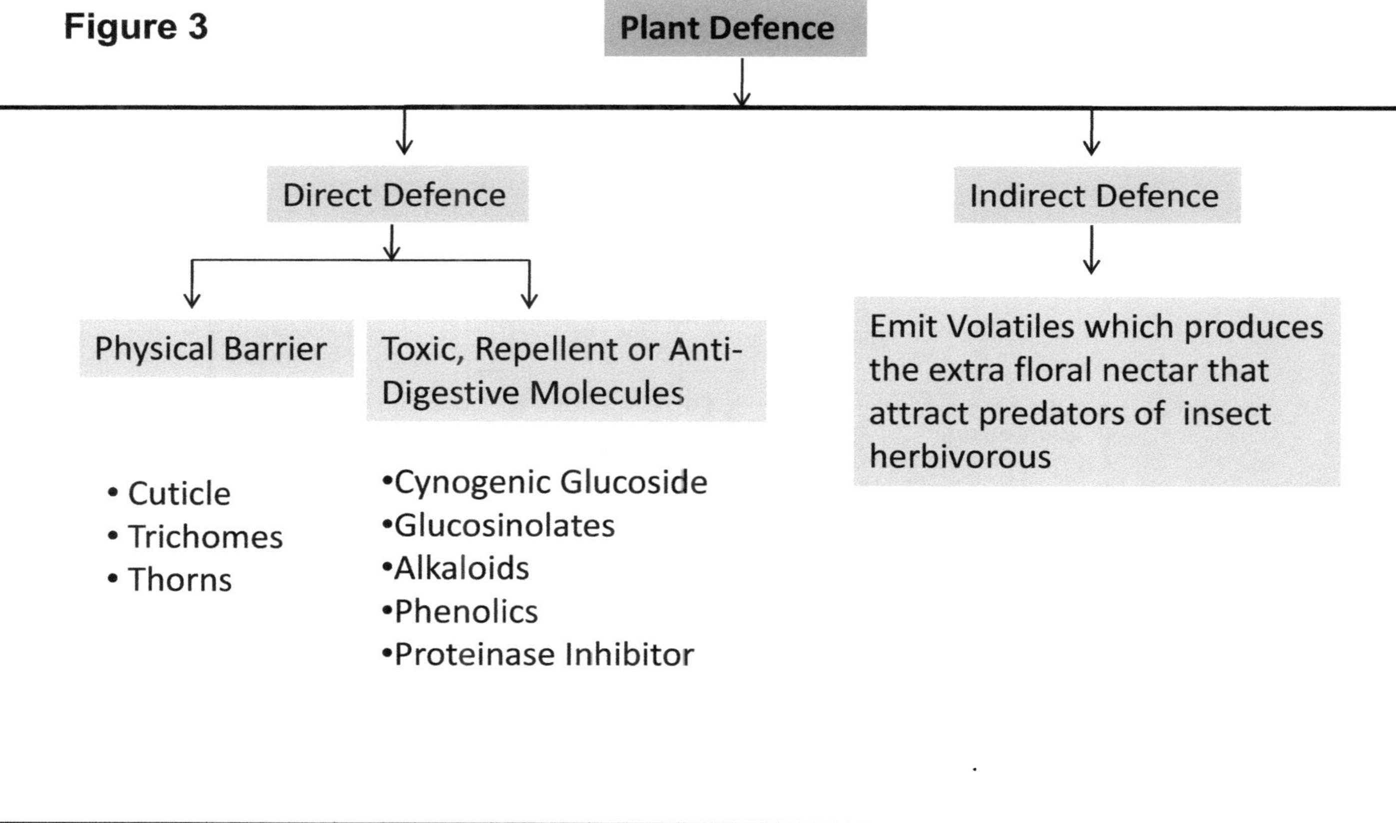

Figure 4

Wounding (Herbivory)
Insect derived elicitor
Local peptide hormone release
Physical Damage
Prosystemin
Threonine Deaminase
Photolytic cleavage
Ethylene
Systemin
Threonine
Isolucine + α Keto Butarate (Ile)
Octadeconide Pathway
Toxic to insect
JASMONIC ACID (JA)
Volatiles e.g. Green leaf volatiles, C10-C15 terpenoids
JA - Ile
Methyl Jasmonate
Auxin, salicylic Acid (-) Ethylene, ABA (+)
Plant detterant / Attracting predators
Plant- Plant interaction (Systemic induction of PI)
Plant – Plant interaction
EARLY (Signaling) GENE INDUCTION in Vascular Bundle
Cell Wall Polysaccharide
Oligalactouronic acid
Activates NADPH oxidase
Hydrogen Peroxide (H$_2$O$_2$)

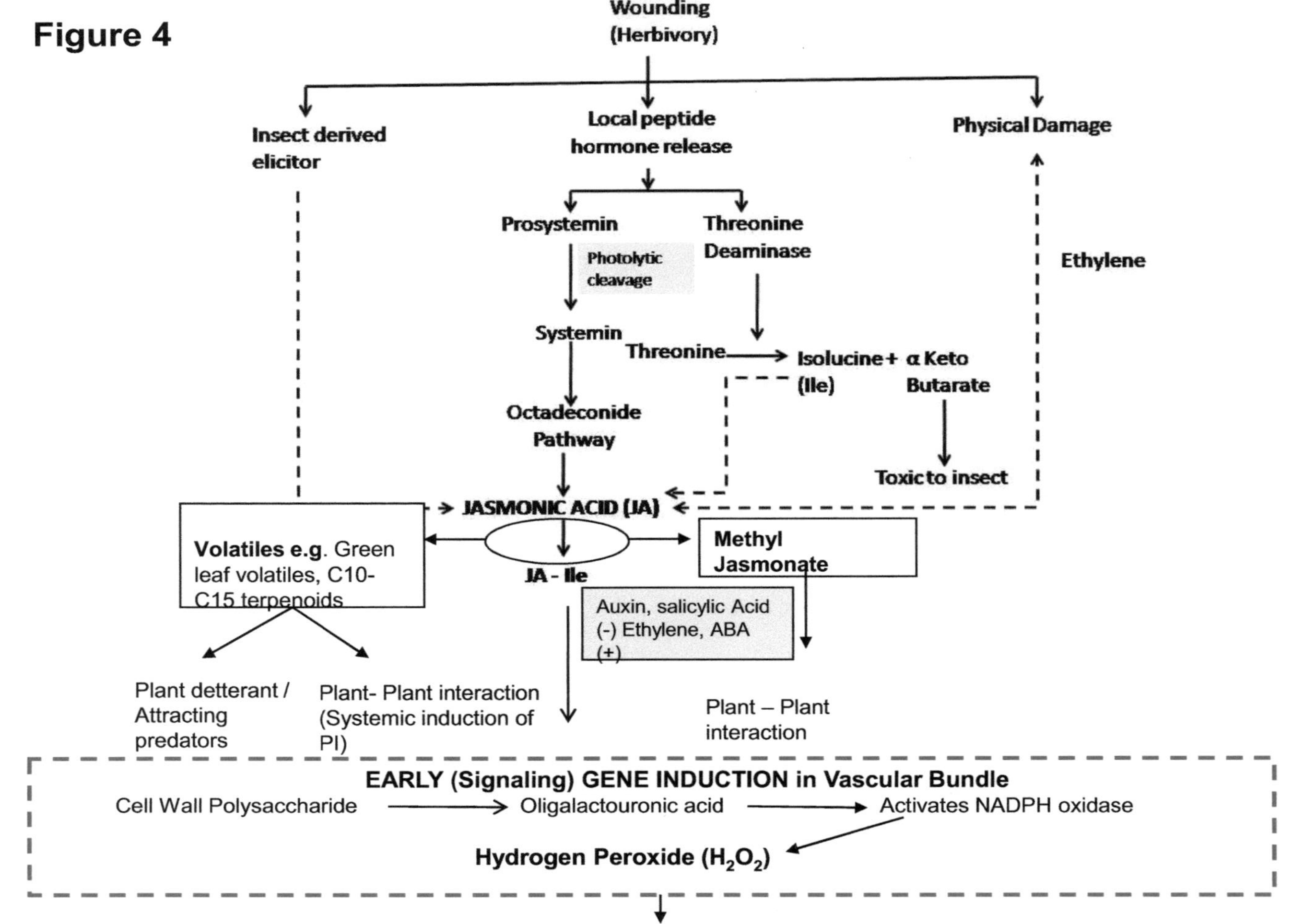

Figure 5

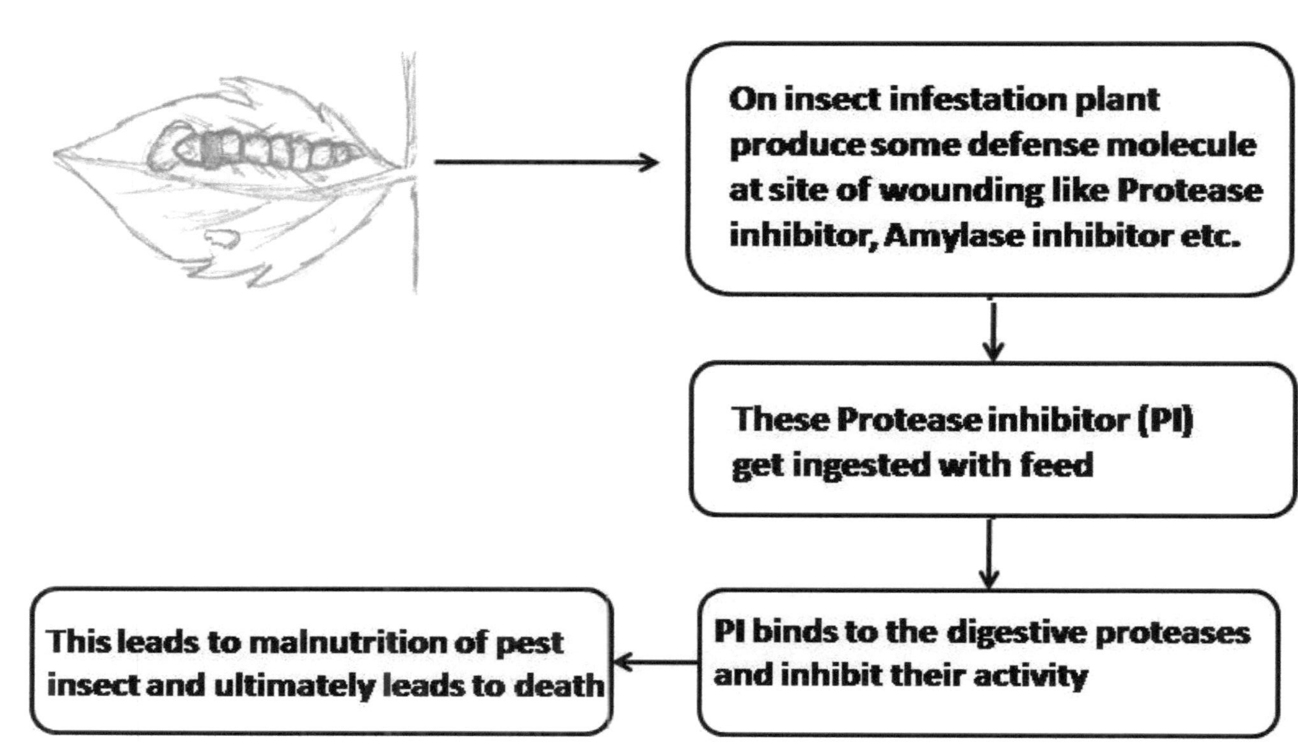

On insect infestation plant produce some defense molecule at site of wounding like Protease inhibitor, Amylase inhibitor etc.
These Protease inhibitor (PI) get ingested with feed
PI binds to the digestive proteases and inhibit their activity
This leads to malnutrition of pest insect and ultimately leads to death

Figure 6

Proteinase Inhibitor
Classes
Serine Protease Inhibitor
Cysteine Protease Inhibitor
Metalloprotease Inhibitor
Asparatyl Protease Inhibitor
Families
•Bowman Birk type
•Kunitz type
•Squash type
•Serpin type
•Mustard seed type
•Ragi seed type
•Wound Inducible Inhibitor
1. Potato Inhibitor Type 1
2. Potato Inhibitor Type 2

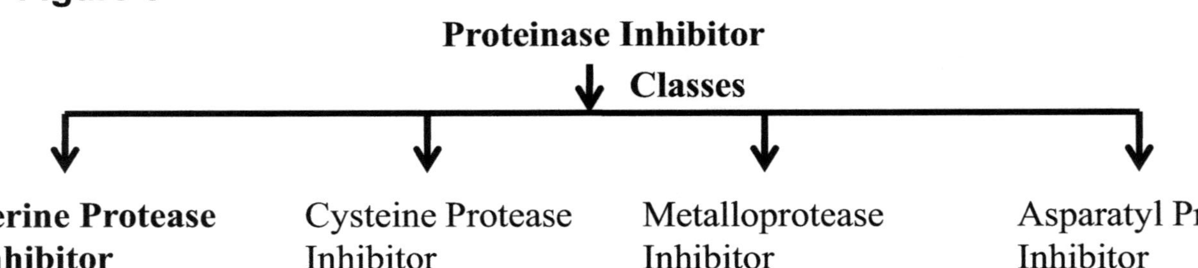

Figure 7

CanPI-7
Signal Peptide
Inhibitory repeat domains (IRD)
1
2
3
4
Linker
Precursor protease inhibitor protein
[1 IRD]
[2 IRD]
[3 IRD]
[4 IRD]
1074,76
6107
12070
12470
19463
25490
2610
26000
Proteolytic processing
Proteolytic processing
5767
5793
1089
Inhibitory repeat domains (IRD)
1
2
3
4

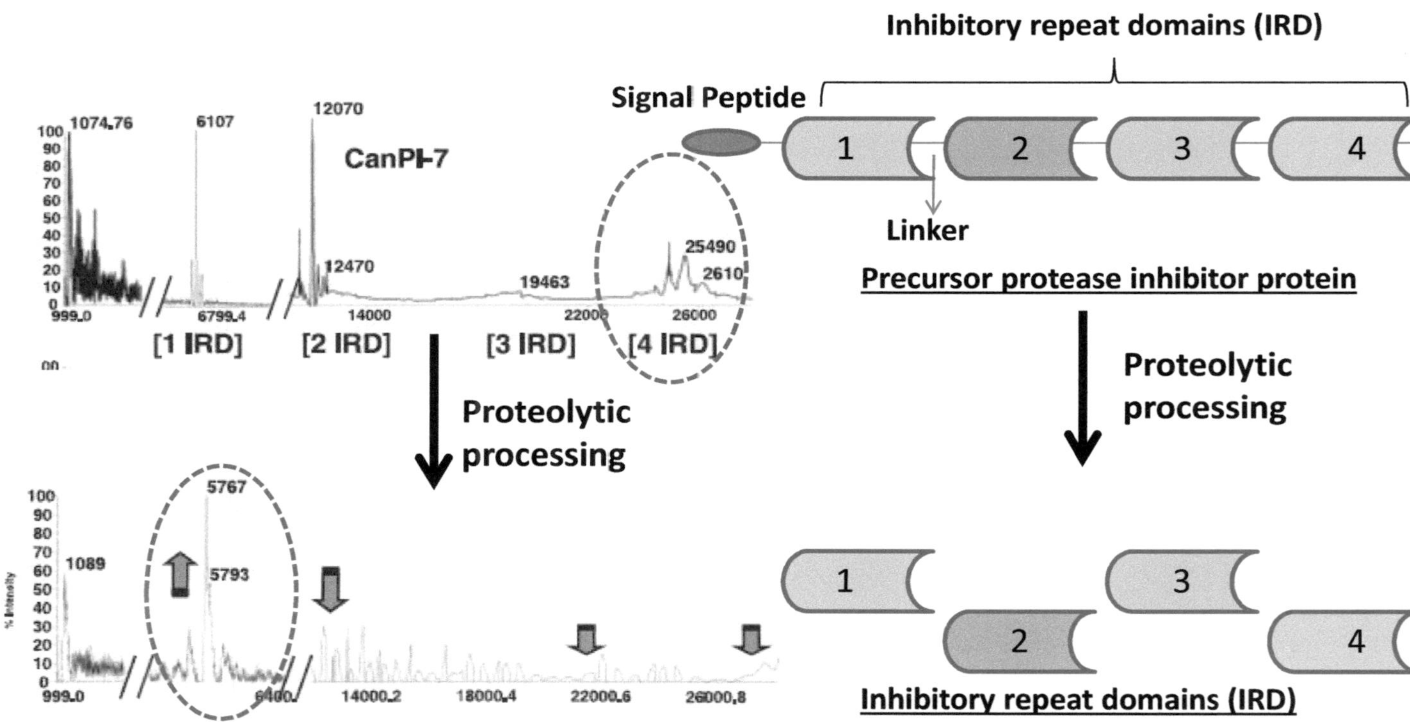